1 **DAY ONE**

Stan, Tony, Jen, and Lori each chose a number.

218 321 208 314

- Jen chose the greatest number.
- Tony chose a number with 3 hundreds.
- Lori did not choose the least number.

What number did each person choose?

2 DAY TWO

Sue is thinking of a 3-digit number.

- It is between 200 and 250.
- Each digit is an even number.
- None of the digits are the same.
- The last two digits have a sum of 10.

What number is Sue thinking of?

Name ______________________________

3 DAY THREE

1. Edgar has 5 coins in his pocket. The total value of the coins is 46¢. What coins does Edgar have?

2. Jean has 8 coins in her pocket. The total value of the coins is 43¢. What coins does Jean have?

Name ______________________________ WEEK 1

4 **DAY FOUR**

Draw the next two terms in the pattern. Explain why they are the next two terms.

5 DAY FIVE

1. How could you get from ☆ to Point 1, then from 1 to 2, and so on, until you reach Point 6? Use the words **right**, **left**, **up**, and **down**.

2. Draw a picture of your path.

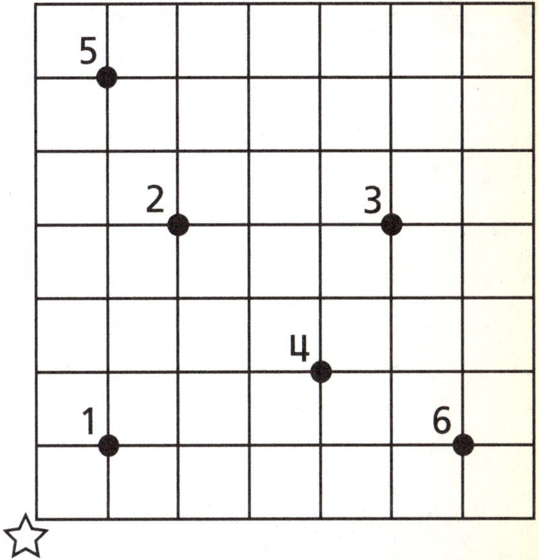

Name ______________________________ WEEK **2**

6 **DAY ONE**

1. Draw a closed figure with a perimeter of 10 units.
2. Draw a closed figure with a perimeter of 20 units.

Name ______________________________

7 **DAY TWO**

Greg has 1 quarter, 2 dimes, and 1 nickel. He puts 3 coins in his piggy bank. What are all the possible amounts of money he could have put in his bank?

Name ______________________________ WEEK **2**

8 **DAY THREE**

Roy spent 20 minutes washing the dog, 60 minutes walking the dog, and 35 minutes training the dog.

1. Did Roy spend more or less than 2 hours with the dog?

2. How do you know?

Name ______________________________

9 **DAY FOUR**

Elena and Rhianne did chores for a total of 75 minutes. Elena did chores for twice as long as Rhianne. For how many minutes did each girl do chores?

Name ____________________

10 **DAY FIVE**

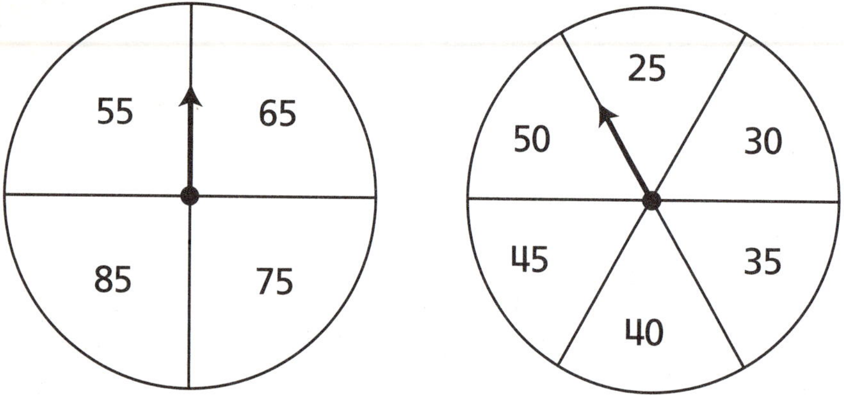

In this spinner game, you spin both spinners and add the two numbers. You get a point if the sum is even. Your friend gets a point if the sum is odd. Is this game fair? Why or why not?

Name ____________________

11 **DAY ONE**

1. What number has:

- a 1 in the hundreds place
- a 7 in the ten thousands place
- a 4 in the thousands and tens places
- a 3 in the ones place?

2. What number is 100 less than that number?

Name ______________________________ WEEK **3**

12 **DAY TWO**

- Cynthia has 2 quarters, 1 dime, 2 nickels, and 4 pennies.
- Sheila has 3 quarters.
- Tracey has 6 nickels, 4 dimes, and 4 pennies.

1. Who has the most money?

2. How much money does each girl have?

Name ______________________________

13 **DAY THREE**

1. Draw a closed figure with an area of 9 square units.

2. Draw a closed figure with an area of 25 square units.

Name ______________________________ **WEEK 3**

14 **DAY FOUR**

This table shows the favorite colors of Mr. Black's students.

Color	**Tallies**
Red	𝍸 𝍷𝍷𝍷𝍷
Blue	𝍸
Green	𝍸 𝍷𝍷𝍷
Yellow	𝍷𝍷

1. How many students chose each color?

2. Make a bar graph that shows the data in the table.

15 **DAY FIVE**

Find each missing number. Explain how you found each number.

1. $9 + \square = 27$
2. $\square + 32 = 71$
3. $600 + \square + 8 = 638$
4. $\square + 100 + 20 = 7{,}120$

16 **DAY ONE**

Laura has 2 quarters, 2 dimes, 3 nickels, and 1 penny. She puts 2 coins in her coin jar. What are all the possible amounts of money she could have put in the jar?

17 **DAY TWO**

Use the numbers in the box to answer the questions. You may use each number only once.

115	137
157	163
193	195

1. Which two numbers have a sum of 300?

2. Which two numbers have a difference of 80?

3. When you add the digits in each number, which two numbers have the same sum?

Name ____________________ WEEK **4**

18 **DAY THREE**

Luis got home at 5:35. It took him 20 minutes to ride his bike home from Ian's house. He was at Ian's house for 2 hours 45 minutes. Before that, Luis was at the library for 1 hour. It took Luis 10 minutes to ride from the library to Ian's house. It took him 15 minutes to ride his bike to the library from his own house. What time did Luis leave his house?

Name ______________________________

19 **DAY FOUR**

Joelle is deciding what to wear today. She can wear blue jeans or a gray skirt. She can wear a white, black, yellow, or red sweater.

1. How many different outfits can she wear?

2. What are they?

20 **DAY FIVE**

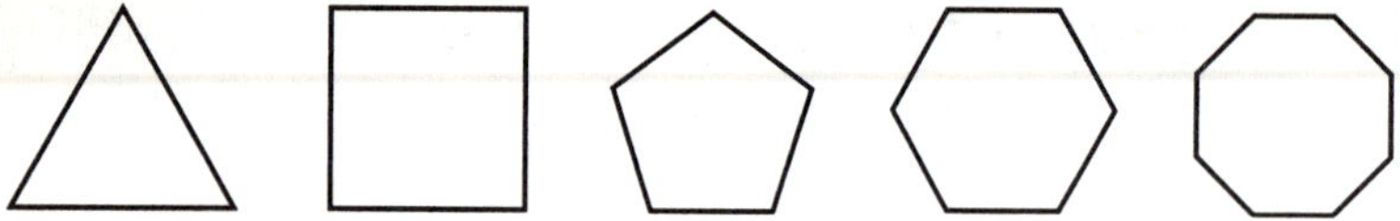

Name three ways these figures are alike.

21 **DAY ONE**

Dave buys lemonade at the snack bar. He gives the cashier $2. The cashier gives Dave 5 dimes, 6 nickels, and 11 pennies for change.

1. How much did the lemonade cost?
2. If Dave asks the cashier to give him his change using the fewest number of coins, what coins will she give him?

Name ______________________________ WEEK **5**

22 **DAY TWO**

Match each multiplication fact with its answer.

1. 2×3	**a.** 6
2. 4×7	**b.** 18
3. 9×2	**c.** 24
4. 7×5	**d.** 28
5. 6×6	**e.** 35
6. 8×3	**f.** 36

23 **DAY THREE**

Use the numbers in the box to answer the questions. You may use each number more than once.

11	16	21
32	46	
54	67	71
82	90	

1. Which three numbers have a sum of 103?
2. There are four pairs of numbers that have a sum that rounds to 90. What are they?
3. Which number has twice as many tens as it has ones?

24 DAY FOUR

Make an 8–by–8 grid like this one. Label START on your grid in the lower left corner. Follow the directions in order. Draw your path, making a dot on the grid after each set of directions. Where do you end?

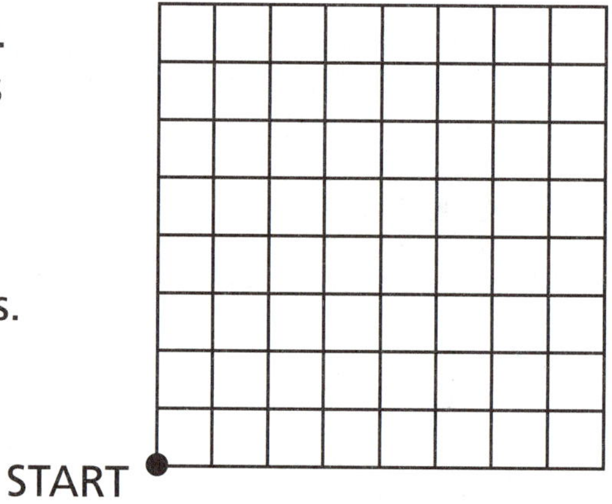

Directions:

- right 2, up 1
- left 1, up 2
- left 1, up 4
- right 3, up 1
- right 2, down 3
- right 3, down 2
- left 4, up 3
- left 1, down 5
- right 5, down 1

25 **DAY FIVE**

Erika was born on February 7, 1992. How old is she on February 1, 2003? In what year does she become 15 years old?

26 **DAY ONE**

$H \times J = 42$

$J \times K = 63$

$54 \div K = H$

$L \times H = 24$

$28 \div L = J$

What is the value of each letter?

Name ______________________________

27 **DAY TWO**

Find each answer. Underline the answers that are greater than 208.

123	98	300	592
+ 77	+ 119	− 116	− 374

28 **DAY THREE**

Cedric says that a square is always a rectangle, but a rectangle is not always a square. Is he right? Why or why not?

29 DAY FOUR

Jackie ran the race in 7 minutes, Craig ran it in 6 minutes, Barbara ran it in 8 minutes, and Carolyn ran it in 7 minutes. Make a bar graph that shows this data.

30 **DAY FIVE**

Match each boy with the distance he rode his bike.

Clues:

- Rob rode farther than Greg but not as far as Cameron.
- Bill rode farther than Cameron.
- Arda rode the shortest distance.

1. Arda	**a.** 5 miles
2. Bill	**b.** 7 miles
3. Cameron	**c.** 10 miles
4. Greg	**d.** 12 miles
5. Rob	**e.** 15 miles

31 **DAY ONE**

The table shows how much money each student has saved this month. Make a pictograph to show this data. Use 25¢ to stand for 25¢.

Name	**Amount**
Andrea	$1
Fred	$3
Jerry	$3
Shar	$2

Name ______________________________

32 DAY TWO

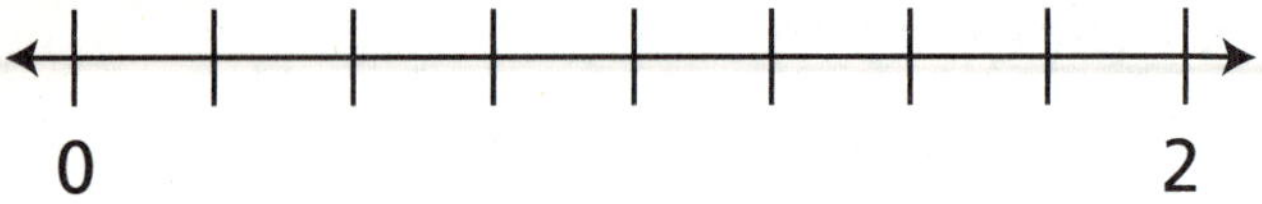

Locate and label $\frac{1}{4}$ and 1 on the number line.

33 **DAY THREE**

Andy is thinking of a 3–digit number.

- Each digit is divisible by 3.
- The tens digit is equal to the sum of the hundreds and ones digits.
- The ones digit is greater than the hundreds digit.

What number is Andy thinking of?

Name ______________________________ WEEK 7

34 **DAY FOUR**

Use these shapes

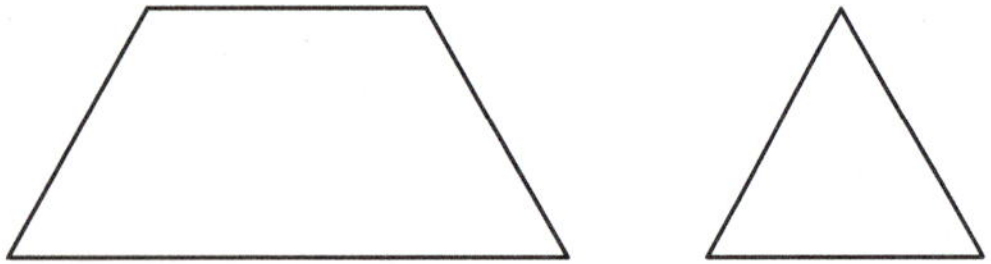

to make this shape.

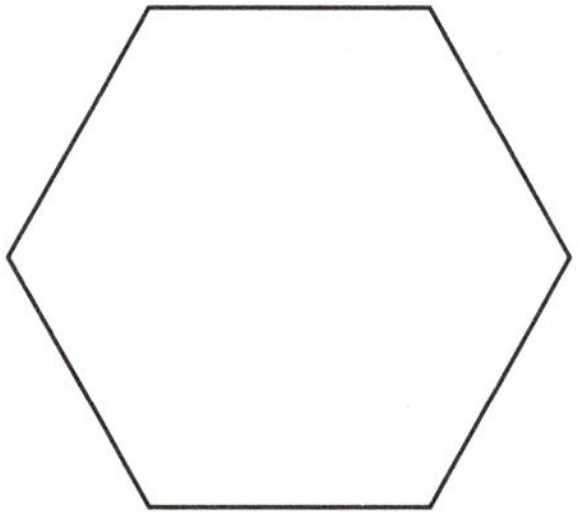

Use a total of 4 smaller shapes to make the larger shape. You may use each shape more than once. Tell how many of each shape you used. Draw a picture of your work.

35 **DAY FIVE**

There are 17 red and 13 blue marbles in a bag. Ms. Ruxton pulls a marble out of the bag without looking.

1. Is Ms. Ruxton more likely to pull out a red or a blue marble? Why?

2. What could Ms. Ruxton do if she wants to have an equally likely chance of pulling out a red or a blue marble?

Name ______________________________

36 **DAY ONE**

Sharon has 9 coins. The total value of the coins is 92¢. Name two ways that Sharon could have 9 coins that equal 92¢.

Name ______________________________

37 **DAY TWO**

Match each division fact with its answer.

1. 81 ÷ 9	**a.** 3
2. 35 ÷ 7	**b.** 4
3. 42 ÷ 6	**c.** 5
4. 27 ÷ 9	**d.** 7
5. 32 ÷ 8	**e.** 9

38 **DAY THREE**

Franco left his house at 9:05 A.M. It took him 15 minutes to get to the doctor's office. He spent 40 minutes there. It took him 5 minutes to get to the grocery store. He shopped for 35 minutes. It took him 20 minutes to get home.

1. What time did Franco get home?

2. How long was he gone?

Name ______________________________

39 **DAY FOUR**

On grid paper, draw a closed figure with a perimeter of 14 units and an area of 9 square units. Explain how you know what the figure's perimeter and area are.

40 **DAY FIVE**

There are 376 first, second, and third graders going to see a play.

- The number of 3rd graders is 28 less than 200.
- The number of 2nd graders is twice the number of 1st graders.

How many students from each grade are going?

Name ______________________________

41 **DAY ONE**

1. What is the greatest 3-digit number you can make using the digits 1–9? Use each digit only once.
2. What is the least 3-digit number you can make? Use each digit only once.
3. What is the difference between the two numbers you made?
4. What is the sum of the two numbers you made?

Name ______________________________

42 **DAY TWO**

Decide whether each is less than 1 foot long, about 1 foot long, or greater than 1 foot long.

- length of a crayon
- height of a piece of paper
- width of your desk
- distance from your teacher's elbow to wrist
- height of a door
- length of a pair of scissors

43 **DAY THREE**

Write >, <, or = in the ◯.

1. 3,000 + 60 + 5 ◯ 3,650

2. seven hundred forty-three ◯ 743

3. 2,908 ◯ two thousand nine hundred eighty

4. 6,795 ◯ 6,000 + 700 + 50 + 9

44 **DAY FOUR**

Put the numbers where they belong in the Venn diagram.

300 500 90 475 325

1,000 903 116 301 527

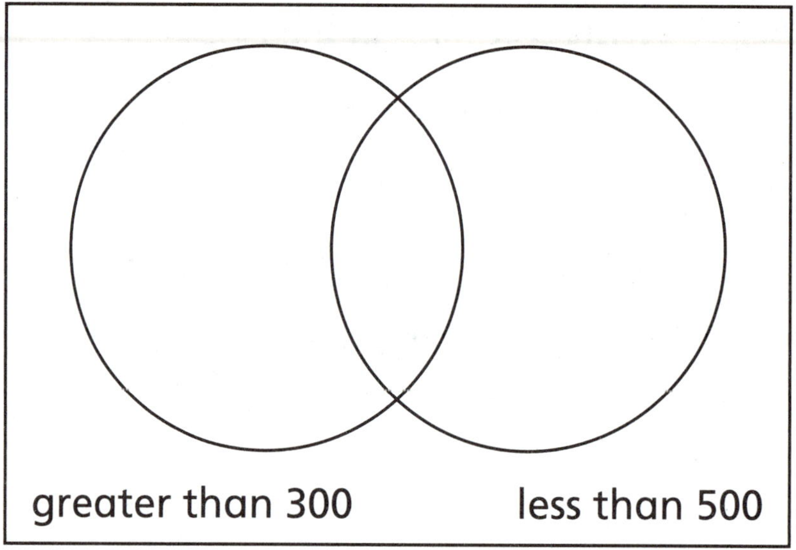

Name ______________________________

45 **DAY FIVE**

Draw the next 3 terms in the pattern and explain why you think they are the correct terms.

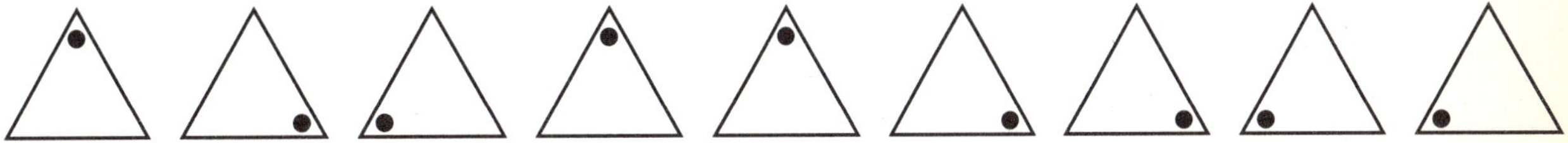

Name ______________________________ WEEK **10**

46 **DAY ONE**

Melissa says that 902 – 768 is 134. Is she right? Prove your answer.

47 **DAY TWO**

Shelly bought 3 packs of 5 pencils. Zoe bought 4 packs of 4 pencils. Who bought more pencils? Draw a picture to show why.

Name ____________________ WEEK **10**

48 **DAY THREE**

Bob and Andy have a 45-minute Math class every school day. Bob says they have Math for 225 minutes each week. Andy says they have Math for 3 hours 45 minutes each week. Who is right? Explain.

Name ______________________________

49 **DAY FOUR**

Tony has 4 quarters, 4 dimes, 4 nickels, and 5 pennies. List three different ways that he can make 65¢.

50 DAY FIVE

Plot the points on a grid. Label each point with its number. Connect the points in order; then connect Point 10 with Point 1. What did you draw?

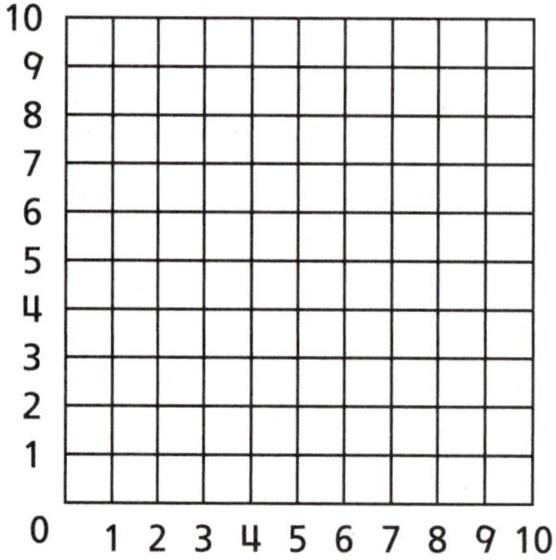

1. (1, 1)
2. (1, 9)
3. (7, 9)
4. (7, 8)
5. (2, 8)
6. (2, 6)
7. (5, 6)
8. (5, 5)
9. (2, 5)
10. (2, 1)

51 **DAY ONE**

There are 4 jars of paint at the art easel. The colors are red, blue, yellow, and white. Pierre mixes two colors together. In how many ways could he mix two colors of paint? What are they?

Name ______________________________

52 **DAY TWO**

What numbers are missing?

1.

$$\begin{array}{r} 7\,\square\,7 \\ +\,\square\,8\,6 \\ \hline 9\,2\,\square \end{array}$$

2.

$$\begin{array}{r} \square\,4\,\square \\ -\,1\,9\,5 \\ \hline 3\,\square\,7 \end{array}$$

53 **DAY THREE**

Write the numbers from 1–100. Then answer these questions:

1. How many times did you write the digit 8?
2. How many numbers have 5 ones?
3. How many 2-digit numbers did you write?

54 **DAY FOUR**

Will has 4 pets. Kermit can fly and is green. The dog is all black. The fish is orange. Loki is a cat. Fluffy does not have legs. Match each pet with its name and its color.

Pets	**Names**	**Colors**
bird	Fluffy	all black
cat	Kermit	black and white
dog	Loki	green
fish	Tosha	orange

Name ______________________________

55 **DAY FIVE**

Put the numbers where they belong in the Venn diagram.

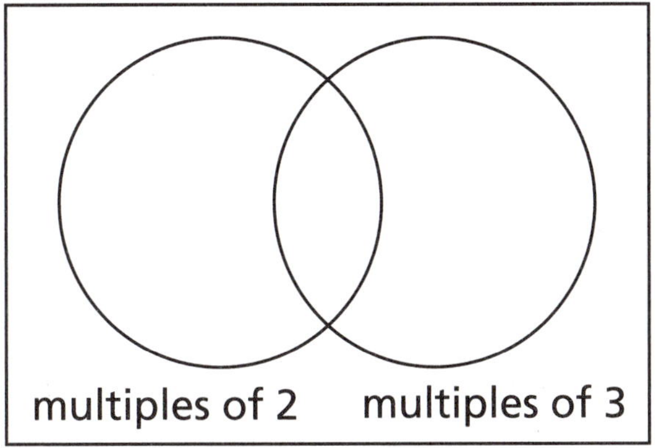

2 3 4 6 8 9 10 12 14 15 16 18 21

56 **DAY ONE**

Jessica says that if you subtract a 3-digit number from a 3-digit number, the answer will be a 3-digit number. Luke says the answer will be a 2-digit number. Mac says the answer will be a 1-digit number. Give examples showing that each student is correct.

57 **DAY TWO**

There are 40 students in the Spanish Club. The students are divided into 3 groups: 1 group meets on Mondays, 1 group meets on Wednesdays, and 1 group meets on Fridays. Twice as many students are in the Monday group as the Wednesday group. The same number of students are in the Wednesday group as the Friday group. How many students are in each group?

58 **DAY THREE**

The clock shows the time that school starts in the morning.

School ends 6 hours 35 minutes later.

1. Draw a picture of what the clock will look like then.

2. Write the time that school ends.

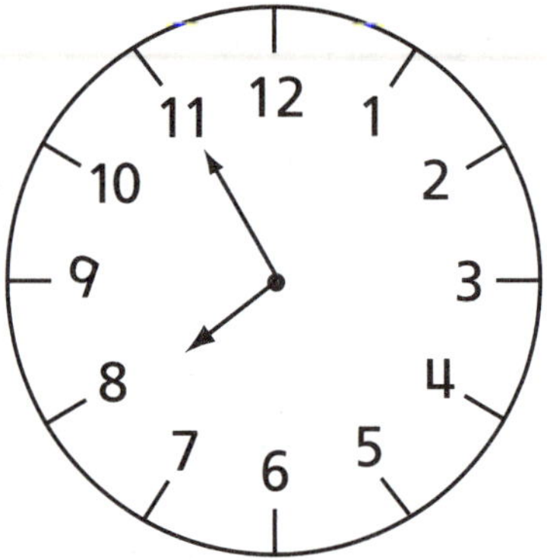

Name ______________________________

59 **DAY FOUR**

The table shows the kinds of birds that Liz saw at her bird feeder last week. Make a horizontal bar graph showing this data.

Type of Bird	Number
Blue Jay	7
Cardinal	5
Crow	10
Magpie	8

Name ______________________________

60 **DAY FIVE**

Jason tosses a 1–6 number cube and spins a 1–4 spinner. He then adds the two numbers together.

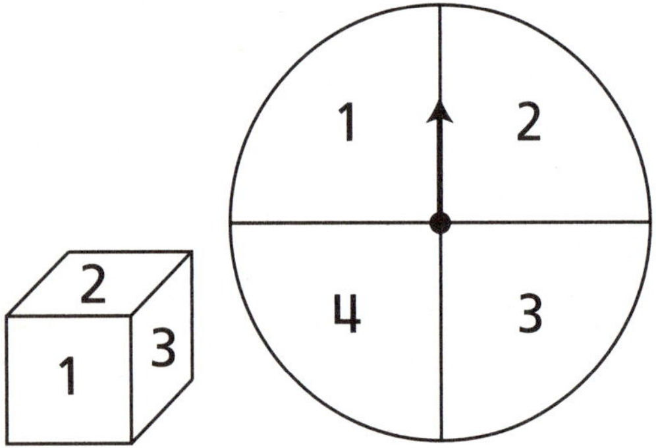

1. Name two outcomes that are impossible.

2. Name two outcomes that are certain.

61 **DAY ONE**

Which of the following figures is congruent to this one?

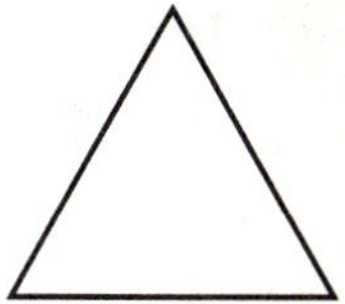

1. 

2. 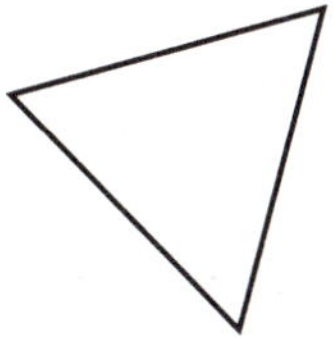

3. 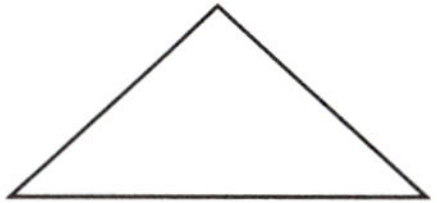

4. 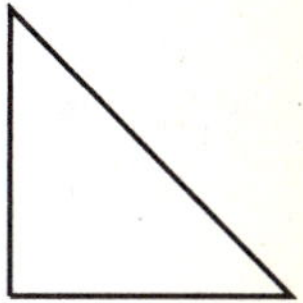

Name ______________________

62 **DAY TWO**

Write each number in standard form. Then write the numbers in order from least to greatest.

1. 3,000 + 400 + 20 + 4
2. three thousand four hundred forty-two
3. 3 thousands, 2 hundreds, 4 tens, 4 ones

63 **DAY THREE**

Draw a number line to show that $3 \times 7 = 21$.
Explain how your number line shows this.

Name ______________________________

 DAY FOUR

1. Complete the table.

Boats	1	2	3	4	5	6	7
Number of People	4	8	12				

2. How many boats would you need to carry 29 people? Explain.

Name ____________________

65 **DAY FIVE**

Bianca paints 3 tiles every 10 minutes. Andrea paints 4 tiles every 15 minutes. Who paints more tiles in 1 hour? How many more tiles does she paint?

66 **DAY ONE**

Use an inch ruler to draw a square with a perimeter of 20 inches. What is the area of the square you drew?

Name ______________________________

67 **DAY TWO**

Find the value of each letter.

$P \times Q = 20$

$Q \times R = 35$

$P \times S = 32$

$S \times T = 72$

68 **DAY THREE**

Write **true** or **false** for each statement.

1. A pencil weighs about 1 pound.
2. A slice of bread weighs about 1 ounce.
3. A football weighs about 1 pound.
4. A television weighs about 1 ounce.

Name ______________________________

69 DAY FOUR

Rick spins these spinners.

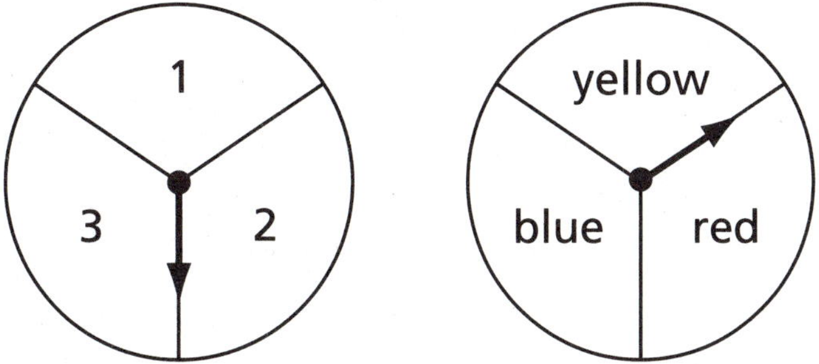

What are all the possible outcomes he could get?

70 **DAY FIVE**

1. What rule could you write for the pattern?
2. Based on this rule, draw what could come next in the pattern.

Name ______________________________

71 DAY ONE

Draw a picture showing why $8 \times 4 = 4 + 4 + 4 + 4 + 4 + 4 + 4 + 4$.
Explain why your picture shows this.

Name ____________________

72 **DAY TWO**

Rosa walked for 1 hour 10 minutes on Saturday, 75 minutes on Sunday, and 1 hour 45 minutes on Monday. How much time did she walk in all? Write your answer as hours and minutes.

Name ______________________________

73 **DAY THREE**

Leo got to school after Emily. Jacques and Kirsten got to school before Emily. Kirsten was not the first one to get to school. In what order did they get to school?

Name ____________________ WEEK **15**

74 **DAY FOUR**

At what times do the hands on a clock form a right angle?

Name ______________________________

75 **DAY FIVE**

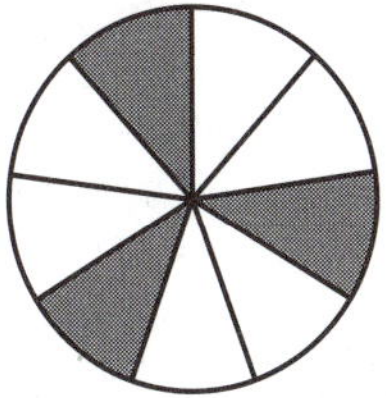

1. What fraction of the circle is shaded?
2. What fraction of the circle is not shaded?
3. What fraction represents the *whole* circle?

Name ______________________________

76 **DAY ONE**

Woody is thinking of a 4-digit number.

- All the digits are multiples of 4.
- The sum of the digits is 24.
- The hundreds and tens digits are two times the thousands and ones digits.

What number is Woody thinking of?

77 DAY TWO

Complete the table.

Tickets	1	2	3	4	5	6	7	8
Price	$0.75	$1.50	$2.25	$3.00				

Name ______________________________ WEEK **16**

78 **DAY THREE**

Miguel bought a book. He paid with a $10 bill and got back a $1 bill, a quarter, a dime, and 3 pennies in change. How much did the book cost?

79 **DAY FOUR**

Our Favorite Books

Key: Each  stands for 3 people.

Which statements about the pictograph are true?

A. Five people chose Comedy.

B. Six people chose Biography.

C. Nine more people chose Comedy than Biography.

D. Eighteen people chose EITHER Drama or Mystery.

80 **DAY FIVE**

Write **slide, flip,** or **turn** to explain how to make the first figure look like the second in each pair. Give a reason for each choice.

1. **2.** **3.**

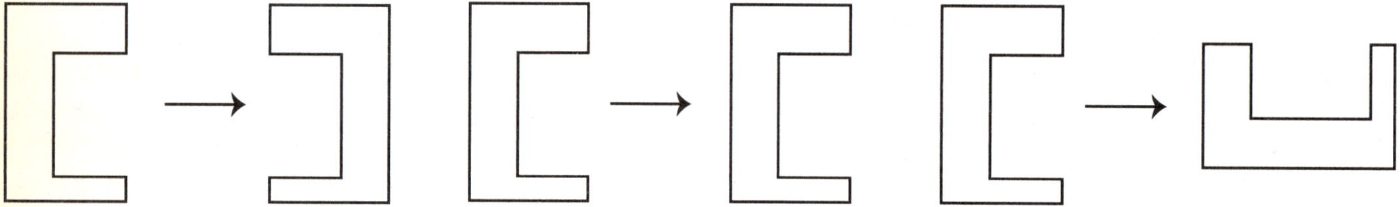

Name ______________________________

81 **DAY ONE**

Draw what comes next in the pattern.

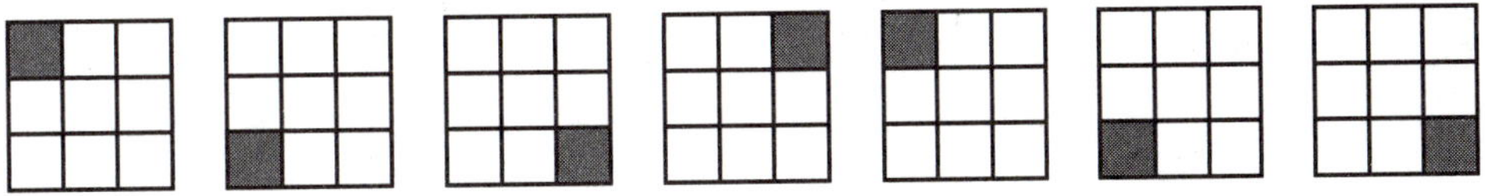

82 **DAY TWO**

Linda has 6 coins in her wallet. All 6 coins are the same. What are all of the possible amounts of money Linda could have?

83 **DAY THREE**

The perimeter of this figure is 91 units. What is the length of the missing side? Explain how you know.

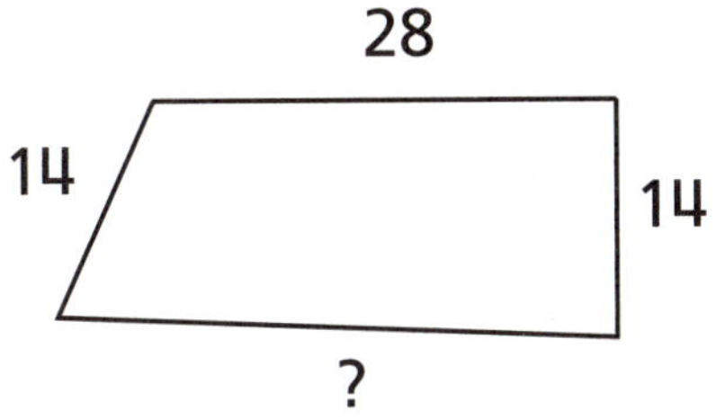

84 **DAY FOUR**

Fill in the missing fractions on each number line.

1. 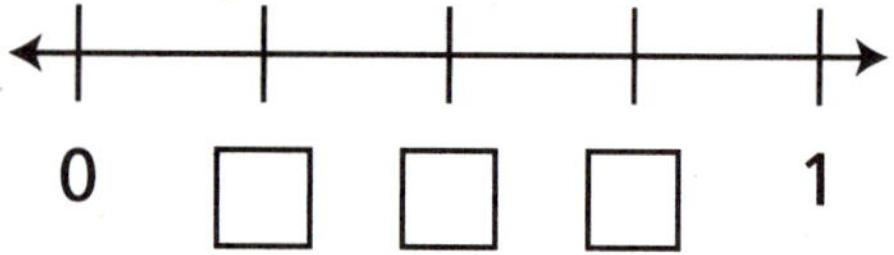

2. 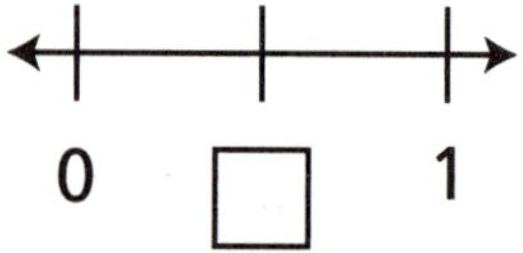

3. 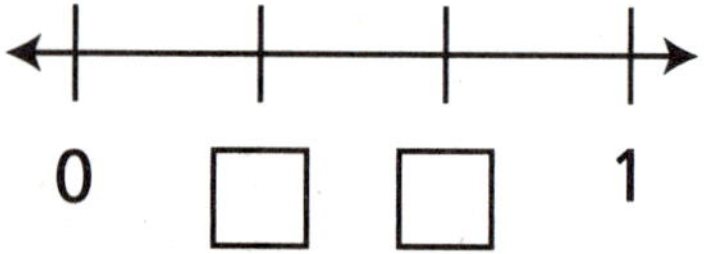

Name ____________________

85 **DAY FIVE**

Four friends are eating lunch. Each is eating and drinking something different from the other three. Mary is not drinking juice or milk. Len is drinking soda. The person drinking orange juice is eating pizza. Steve is eating something that flies. Mary is eating a food made from something that swims. Match each name with his or her food and drink.

Names	**Foods**	**Drinks**
Julie	Chicken	Milk
Len	Hot dog	Orange juice
Mary	Pizza	Soda
Steve	Tuna salad	Water

Name ______________________________ WEEK **18**

86 **DAY ONE**

Gaston buys a shirt for $9.39. He pays with the fewest number of coins and bills possible. What coins and bills does Gaston use?

Name ______________________________

87 **DAY TWO**

Amelia bought 12 apples at the store. $\frac{2}{3}$ of the apples were red and $\frac{1}{3}$ of the apples were green.

1. Draw a picture of the apples Amelia bought.

2. How many red and green apples did she buy?

88 **DAY THREE**

What numbers come next in the pattern? Why do you think they come next?

100, 211, 322, 433, _____, _____

Name ____________________

89 **DAY FOUR**

Give the ordered pair for each point on the grid.

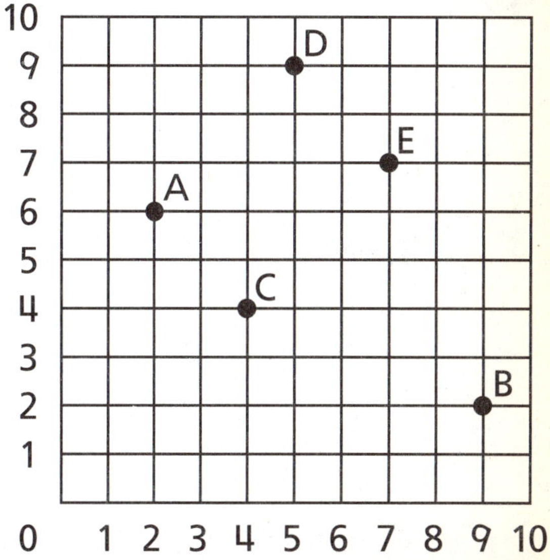

90 DAY FIVE

Use these shapes

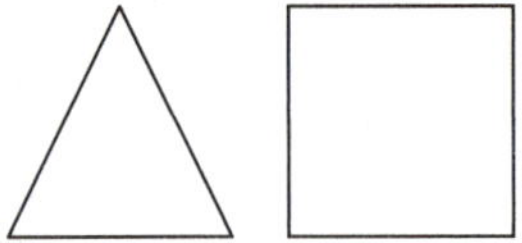

to make this shape.

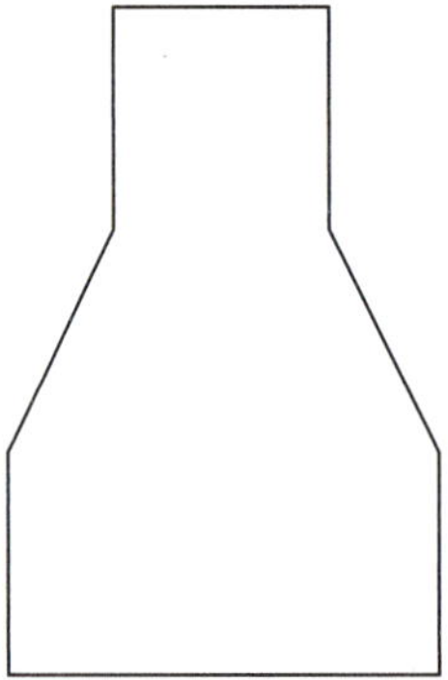

Use a total of 6 smaller shapes to make the larger shape. You may use each shape more than once. Tell how many of each shape you used. Draw a picture of your work.

Name ______________________________

91 **DAY ONE**

Punky can eat wet, moist, or dry cat food. She can drink milk or water. How many different combinations of food and drink can she have? What are they?

Name ______________________________

92 DAY TWO

1. Draw a rectangle. Shade it to show $\frac{1}{3}$.
2. Draw a square. Shade it to show $\frac{3}{8}$.

Name ______________________________

 DAY THREE

Use the numbers in the box to answer the questions. You may use each number more than once.

97	284
398	519
691	983

1. Which number is **about** 100?
2. Which number is **about** 1,000?
3. Which two numbers have a sum of **about** 1,000?
4. Which two numbers have a difference of **about** 100?

94 **DAY FOUR**

Draw a rectangle with a perimeter of 22 inches.

1. How long is each side?

2. What is the area of the rectangle you drew?

95 **DAY FIVE**

The clock shows what time Nancy left home.

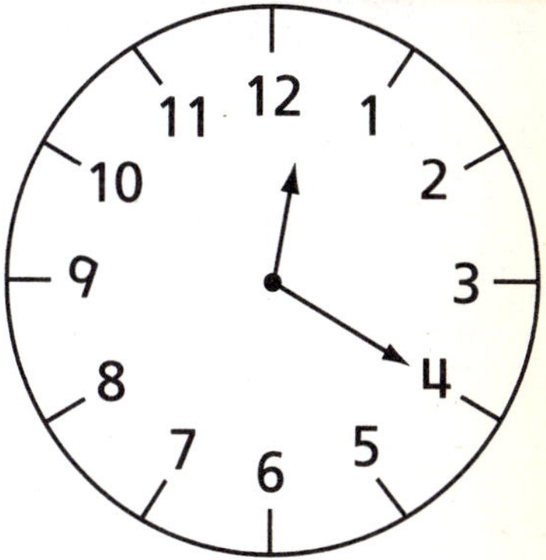

She arrived at swim practice 10 minutes after she left home. Swim practice lasted 2 hours 15 minutes. She arrived at Pizza Palace 25 minutes after swim practice ended. She stayed at Pizza Palace for 45 minutes. She got home 20 minutes after leaving Pizza Palace. What time did Nancy get home?

96 **DAY ONE**

What numbers are missing?

1.

$$\begin{array}{r} 3\ 9\ \square \\ +\ 4\ \square\ 2 \\ \hline \square\ 5\ 3 \end{array}$$

2.

$$\begin{array}{r} \square\ \square\ 9 \\ -\ \ \ 9\ 5 \\ \hline 5\ 1\ \square \end{array}$$

97 **DAY TWO**

Jorge makes a pattern. The rule for his pattern is **slide**, **flip**, **slide**, **flip**, and so on. He starts with this shape:

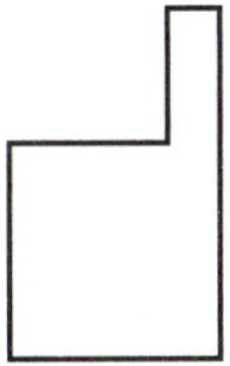

Draw the first 6 shapes in Jorge's pattern.

Name ____________________ WEEK **20**

98 **DAY THREE**

There are 5 cars, 2 tricycles and 4 bicycles parked in the driveway. How many wheels is that?

Name ____________________

99 **DAY FOUR**

Cindy has these number cards: 2 0 7 4

Brad has these number cards: 6 1 3 8

1. Who can make the greatest 4-digit number with the number cards? What is it?
2. Who can make the least 4-digit number with the number cards? What is it?

Name ____________________

100 **DAY FIVE**

The table shows the snacks Ms. Dehmel's class brought for the party. Make a bar graph that shows this data.

Item	**Number of People**
Chips	9
Cookies	4
Fruit	6
Popcorn	7

Name ______________________________

101 **DAY ONE**

Find each sum or difference. Underline the answer that is greatest. Circle the answer that is least.

1. 769 + 254

2. 248 + 653

3. 586 − 437

4. 704 − 169

Name ____

102 **DAY TWO**

Toss a coin 20 times. Make a tally table to record the number of times it lands on "heads" and "tails." Then make a bar graph to show the data.

103 **DAY THREE**

What is the pattern? What are the next three numbers in the pattern?

195, 182, 169, 156, 143, _____, _____, _____

Name ______________________________

104 **DAY FOUR**

Which statements are reasonable?

A. A pen is about 14 centimeters long.

B. Your foot is about 50 centimeters long.

C. A cat is about 10 centimeters tall.

D. A can of soda is about 12 centimeters tall.

Name ______________________________

105 **DAY FIVE**

Draw a rectangle that has an area of 21 square centimeters. What is its perimeter?

Name ______________________ **WEEK 22**

106 **DAY ONE**

Nicole has $1.05 in coins. She has 6 coins. What coins could she have?

Name ______________________________

107 **DAY TWO**

Ricky read 20 pages on Monday; 15 pages each day on Tuesday, Wednesday, and Thursday; and 25 pages on Friday. How many pages did he read in all? Show how you found your answer.

108 **DAY THREE**

Tabitha and Moe take turns spinning the spinner. Tabitha gets a point if the spinner lands on an open figure. Moe gets a point if the spinner lands on a closed figure. Is this fair? Why or why not?

Name ____________________

109 **DAY FOUR**

Draw a right triangle with two sides that are 5 inches long. About how many inches long is the third side?

Name ________________________________ WEEK **22**

110 **DAY FIVE**

Match each decimal with its equivalent fraction.

1. 0.01 **a.** $\frac{1}{10}$

2. 0.07 **b.** $\frac{7}{10}$

3. 0.1 **c.** $\frac{1}{100}$

4. 0.7 **d.** $\frac{7}{100}$

Name ____

111 **DAY ONE**

Mr. Fox cut out 84 shapes. He cut out twice as many squares as circles. He cut out twice as many triangles as squares. He didn't cut out any other shapes. How many of each shape did he cut out?

Name ______________________________ WEEK **23**

112 **DAY TWO**

Find the value of each letter.

$81 \div J = J$

$27 \div J = K$

$18 \div L = K$

$24 \div M = L$

Name ________________________________

113 DAY THREE

SoHee lives 25 minutes from the library. She left her house to go to the library at 2:45 P.M. and got home at 5:30 P.M. How long was SoHee at the library?

114 **DAY FOUR**

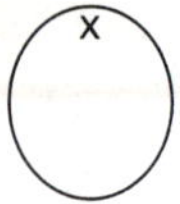 = brown hair

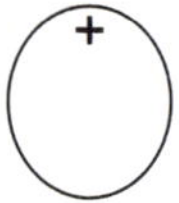 = black hair

 = blonde hair

 = red hair

 = brown eyes

 = green eyes

 = blue eyes

 = girl

 = boy

What does  mean?

Name ______________________________

115 **DAY FIVE**

Complete the table. How many plants are in 8 rows?

Rows	1	2	3	4	5	6	7	8
Plants	7	14	21	28				

Name ______________________ WEEK **24**

116 **DAY ONE**

Write the answers in order from greatest to least.

1. $92 - 51$
2. 8×3
3. $39 + 45$
4. $45 \div 5$

Name ______________________________

117 **DAY TWO**

Mark's printer prints 3 pages each minute. How long will it take for him to print a 15-page report? a 27-page report?

Name ______________________________ WEEK **24**

118 **DAY THREE**

Explain what is wrong with the number line. Then draw the number line so it is correct.

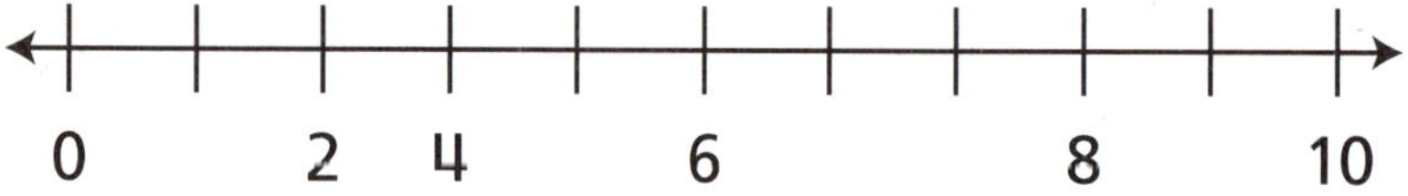

119 **DAY FOUR**

Li makes a pattern. The rule for the pattern is **flip**, **turn**, **flip**, **turn**, and so on.

She starts with this shape:

Draw the first 6 shapes in Li's pattern.

120 **DAY FIVE**

Which shape is **not** a hexagon? Why?

1. 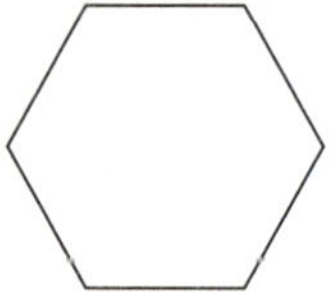

2. 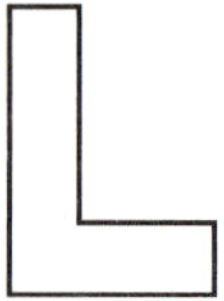

3. 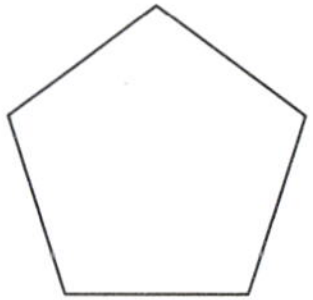

4. 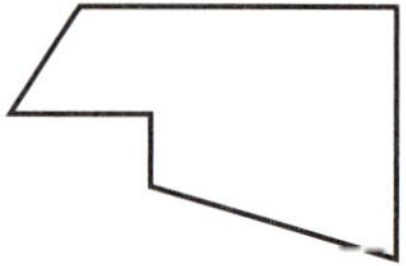

Name ______________________

121 **DAY ONE**

Tim spins the two spinners. What are all the possible products he could get?

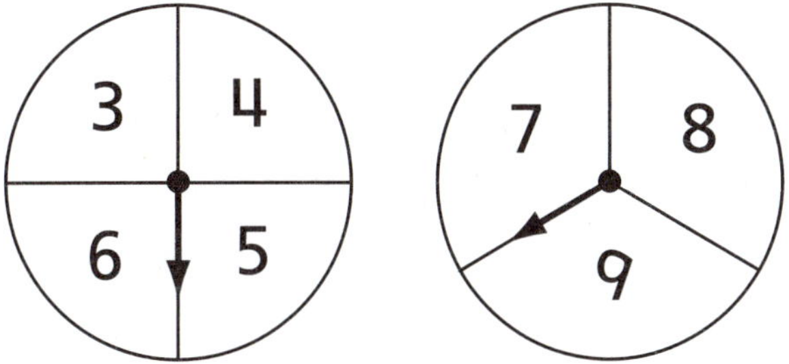

122 **DAY TWO**

Marla is thinking of a 4-digit number.

- It has no ones.
- The number of thousands is 2×4.
- The number of hundreds is 5 less than the number of thousands and 4 less than the number of tens.

What number is Marla thinking of?

123 **DAY THREE**

Draw 3 different rectangles with perimeters of 14 inches.

1. How long are the sides of each rectangle?
2. What is the area of each rectangle?

Name ____________________ WEEK **25**

124 **DAY FOUR**

Write the decimal shown by the shaded part of each picture.

1. 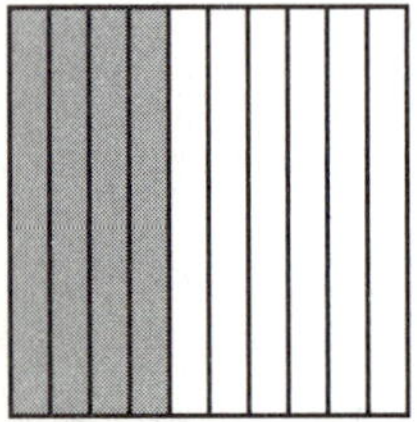

2. 

3. 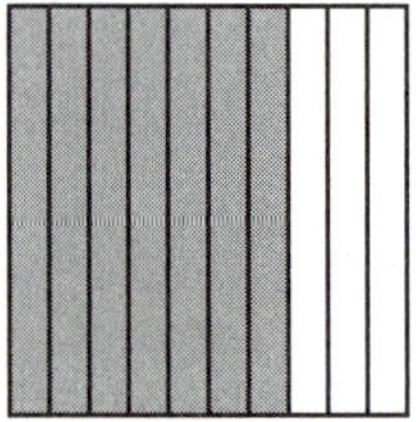

4. 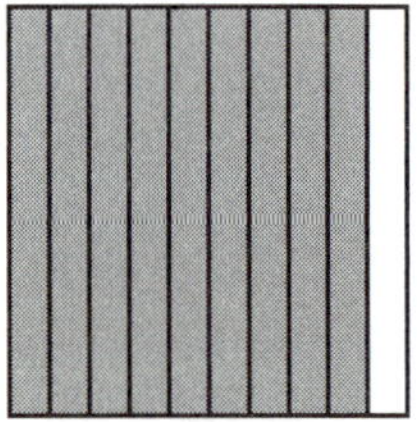

125 **DAY FIVE**

What is the pattern? What are the next three numbers in the pattern?

900, 865, 830, 795, 760, _____, _____, _____

Name ______________________________

126 **DAY ONE**

1. What is the difference between the highest and lowest temperature?
2. Between which two consecutive days did the greatest difference in temperature occur?
3. Which day was warmest?
4. Which day was coldest?

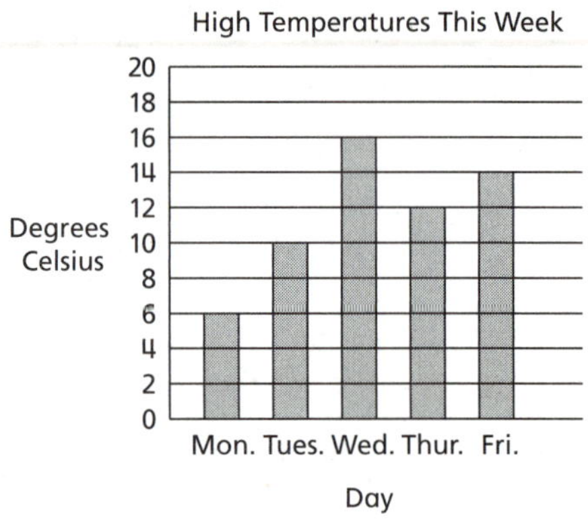

Name ______________________________ WEEK **26**

127 **DAY TWO**

3 hundredths   17 hundredths   8 tenths   1 tenth

1. Write each decimal in standard form.
2. Write the decimals in order from least to greatest.

Name ______________________________

 DAY THREE

Find each missing number.

1. $123 \times \square = 369$

2. $201 \times \square = 804$

3. $\square \times 2 = 868$

129 DAY FOUR

Match each bowl with its capacity.

Clues:

- The white bowl holds more than the blue bowl.
- The red bowl holds the most.
- The green bowl holds less than the blue bowl.

1. blue	**a.** 1 cup
2. green	**b.** 1 gallon
3. red	**c.** 1 pint
4. white	**d.** 1 quart

130 **DAY FIVE**

How could you use these shapes

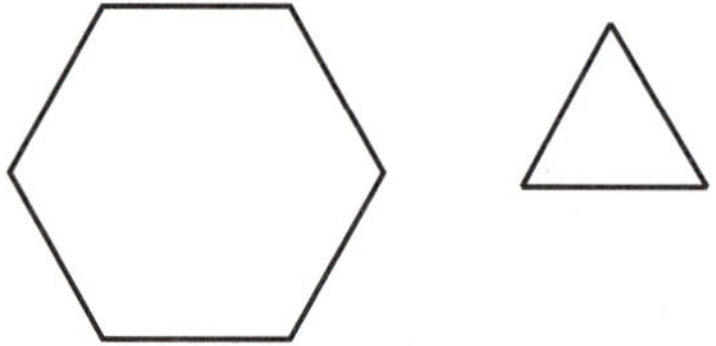

to make this shape?

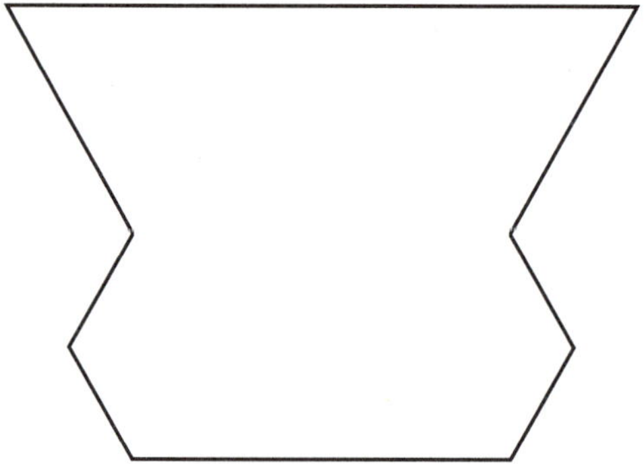

Use a total of 10 smaller shapes to make the larger shape. Tell how many of each shape you used. Draw a picture of your work.

Name ______________________________

131 DAY ONE

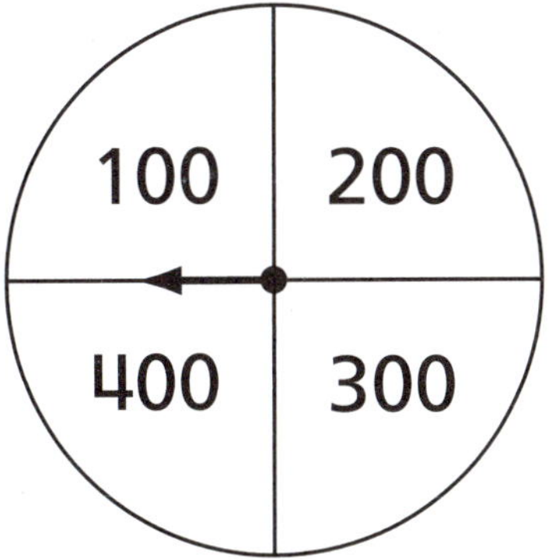

What is the least number of spins needed to get a sum of 1,000? What numbers must the spinner land on to get 1,000 in the least number of spins?

132 **DAY TWO**

The quotient is 18 R3. The divisor is 4. What is the dividend? How do you know?

Name ______________________________

133 **DAY THREE**

I am a 4-sided figure. All my sides are the same length. I am **not** a square. What am I?

134 **DAY FOUR**

Complete the table.

Rule: Multiply by 5	
Input	**Output**
5	25
7	
12	
25	

Name ______________________________

135 DAY FIVE

Wendy went to the park for 3 hours 35 minutes. Norm went for 220 minutes. Who went to the park for longer? Tell how you know.

136 **DAY ONE**

Put the numbers where they belong in the Venn diagram.

0.51   0.37   0.2   0.7   0.15   0.06

0.4   0.13   0.09   0.62   0.18

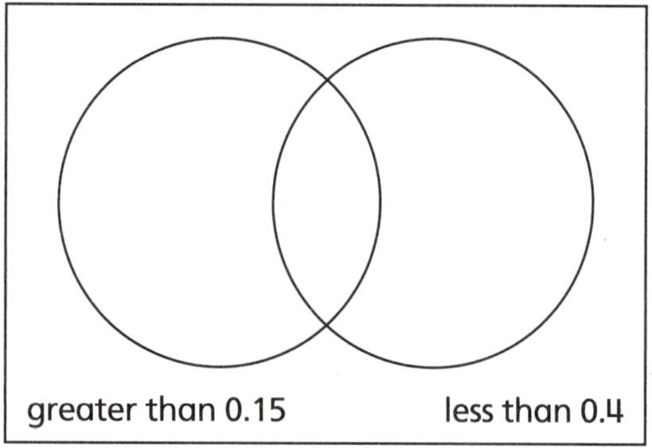

Name ______________________________

137 **DAY TWO**

The rule for Alex's pattern is **right angle, line segment, triangle, right angle, line segment, triangle,** and so on. Draw the first 8 shapes in Alex's pattern.

Name ______________________________ WEEK **28**

138 **DAY THREE**

Follow the directions. Give the answer to each step.

- Start: 201
- Add 195
- Subtract 301
- Divide by 5
- Multiply by 7
- Subtract 130

Name ______________________________

139 **DAY FOUR**

Match the equivalent fractions.

1.	$\frac{1}{2}$	**a.**	$\frac{2}{6}$
2.	$\frac{1}{3}$	**b.**	$\frac{2}{12}$
3.	$\frac{1}{4}$	**c.**	$\frac{3}{6}$
4.	$\frac{1}{6}$	**d.**	$\frac{3}{24}$
5.	$\frac{1}{8}$	**e.**	$\frac{3}{12}$

Name ______________________________

140 **DAY FIVE**

Shade each picture to show the given fraction.

1. 

$\frac{1}{2}$

2. 

$\frac{2}{3}$

3. 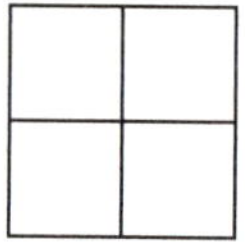

$\frac{3}{4}$

4. 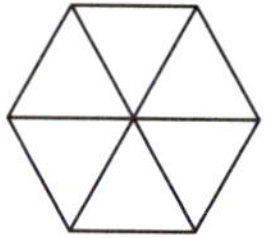

$\frac{1}{3}$

Name ______________________________

141 **DAY ONE**

What numbers are missing?

1.

$$\begin{array}{r} 4\square 3 \\ \times \quad \square \\ \hline \square 2\,6 \end{array}$$

2.

$$\begin{array}{r} \square 1 \\ \square \overline{)\,9\,3} \end{array}$$

142 **DAY TWO**

In art class you can make a picture with paint, charcoal, crayons, or pencil. You can choose white, yellow, or blue paper. How many different ways can you make a picture? What are they?

Name ______________________________

143 DAY THREE

Fill in the missing numbers on the number line.

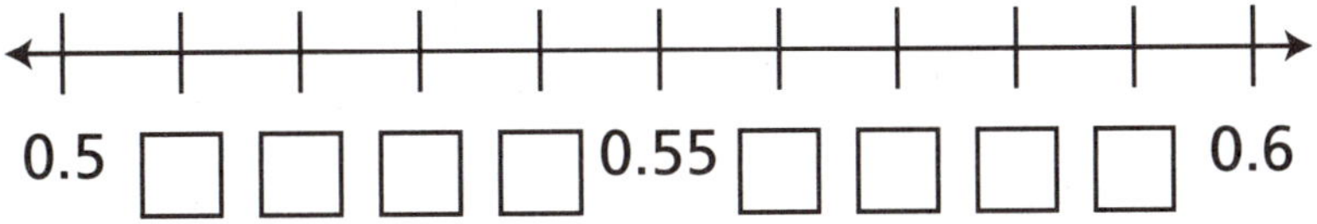

144 **DAY FOUR**

Match each solid figure with its name.

1. 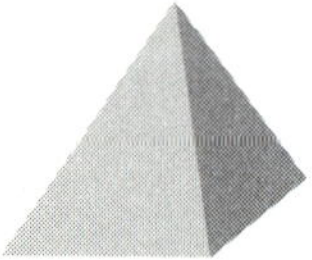

2. 

3. 

4. 

5. 

6. 

a. cone
b. cube
c. cylinder
d. prism
e. pyramid
f. sphere

Name ______________________________

145 DAY FIVE

Carmen has these number cards:

12	1	1	11	7	8	12	6	7	11

She mixes them up and places them face down on the table. She chooses a card without looking.

1. What are all the possible outcomes Carmen could get?

2. Name one outcome that is certain.

3. Name one outcome that is impossible.

Name ______________________________ WEEK **30**

146 **DAY ONE**

Write a story problem that can be solved by multiplying 3 and 24. Solve your story problem.

147 **DAY TWO**

Draw a square that is 10 centimeters on each side. Draw a triangle above the square so it looks like a house. Draw two square windows that are 1 centimeter on each side inside the house. Draw a door that is 3 centimeters wide and 5 centimeters tall.

Name ______________________________ WEEK **30**

148 **DAY THREE**

Draw two congruent triangles.

Name ____________________

149 **DAY FOUR**

The perimeter of the rectangle is 26 centimeters. The area of the rectangle is 42 square centimeters. What is the length of the longer side?

6 cm

?

150 **DAY FIVE**

Give a rule for the pattern. Then draw the next three shapes in the pattern.

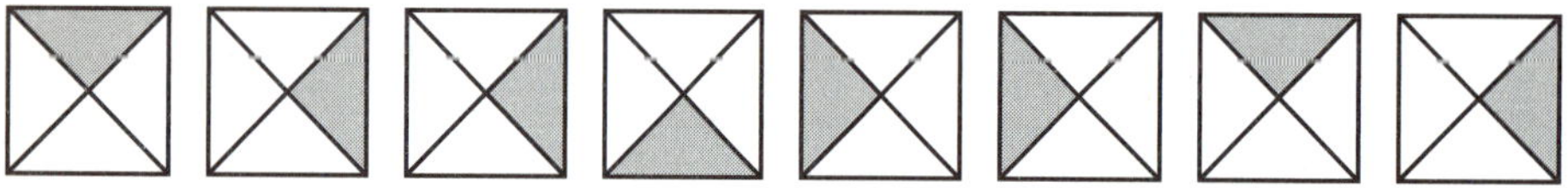

Name ______________________________

151 **DAY ONE**

Manuel has 2 one-dollar bills, 3 quarters, 2 dimes, and 1 nickel. He gives 1 bill and 2 coins to his sister. How much money could he have given her?

Name ______________________________

152 **DAY TWO**

Find each difference. Write the answers in order from greatest to least.

1. $\begin{array}{r} 975 \\ -\ 100 \\ \hline \end{array}$

2. $\begin{array}{r} 975 \\ -\ 10 \\ \hline \end{array}$

3. $\begin{array}{r} 608 \\ -\ 100 \\ \hline \end{array}$

4. $\begin{array}{r} 608 \\ -\ 10 \\ \hline \end{array}$

153 **DAY THREE**

Write a story problem that can be solved by dividing 68 by 4. Solve your story problem.

154 **DAY FOUR**

1. Which capital letters of the alphabet do **not** have a line of symmetry?
2. Draw one line of symmetry for each of the letters that does have one.

155 **DAY FIVE**

Plot the points on a grid. Label each point with its number. Connect the points in order from 1–10.

1. (1, 1)

2. (1, 9)

3. (9, 9)

4. (9, 2)

5. (3, 2)

6. (3, 7)

7. (7, 7)

8. (7, 4)

9. (5, 4)

10. (5, 6)

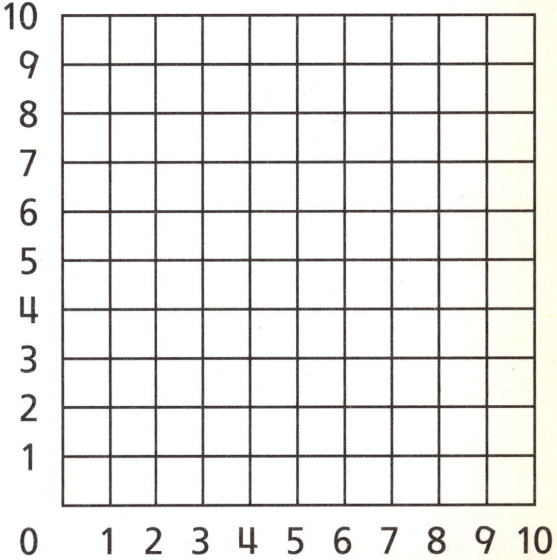

Name ______________________________

156 **DAY ONE**

Estimate the length of each object to the nearest inch.

1. a crayon
2. scissors
3. a paper clip
4. a new pencil

157 **DAY TWO**

Put the numbers in order from least to greatest.

3 tenths 0.29 $\frac{16}{100}$ $\frac{2}{10}$ 4 hundredths

158 **DAY THREE**

Estimate the height of each object to the nearest foot.

1. your desk
2. a cat
3. a door
4. a window

Name ______________________________

159 **DAY FOUR**

There are 25 students in Libby's class. Of those students, $\frac{3}{5}$ ride a bus to school. Draw a picture to find how many students ride a bus to school.

160 **DAY FIVE**

Write the name of each solid figure in the correct circle in the Venn diagram.

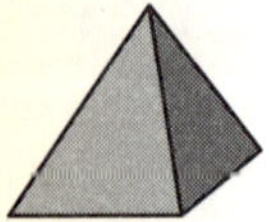
square pyramid

cone

sphere

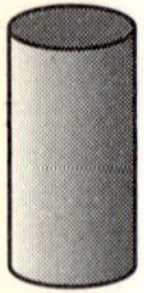
cylinder

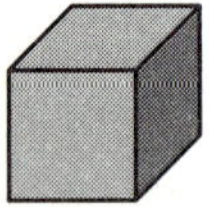
cube

rectangular prism

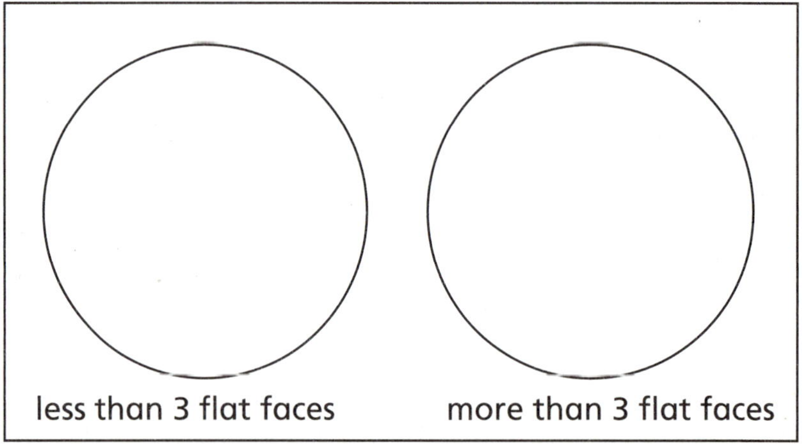

Name ______________________________

161 **DAY ONE**

The Dwyers have 4 dogs.

- Emma is bigger than Mona.
- Farley is not the smallest dog, but is smaller than Emma.
- Sherlock is the biggest dog.

Write the dogs' names in order from smallest to largest.

162 **DAY TWO**

There are 145 second graders, 298 third graders, and 186 fourth graders at the aquarium. Round each number to the nearest hundred to estimate about how many students there are in all.

Name ______________________________

163 **DAY THREE**

Put the fractions in order from least to greatest.

$\frac{3}{6}$ $\frac{2}{3}$ $\frac{3}{4}$ $\frac{5}{12}$

164 **DAY FOUR**

Fill in the blank squares to complete the pattern. Explain your thinking.

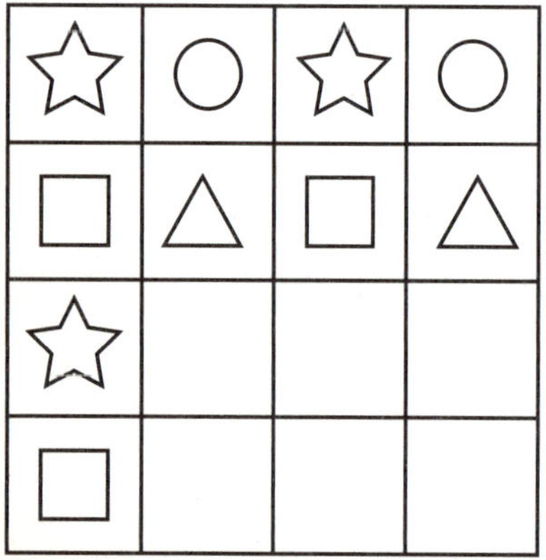

Name ______________________________________ WEEK **33**

165 **DAY FIVE**

Based on the graph, write three statements that are true.

Pets We Have	
Bird	
Cat	
Dog	
Fish	

Key: Each  stands for 4 people.

Name ______________________________

166 DAY ONE

Use the numbers in the box to answer the questions. You may use each number only once.

3	5
9	74
85	98

1. Which two numbers have a product of 222?
2. Which number divided by which number has a quotient of 19 R3?
3. Which two numbers have a difference of 76?

Name ____

167 **DAY TWO**

Draw two different rectangles with areas of 24 square units. What is each rectangle's perimeter?

168 **DAY THREE**

Danielle writes in her journal for 10 minutes every morning. She writes in her journal for 15 minutes every night. How much time does she spend writing in her journal every week? Write your answer as hours and minutes.

Name ______________________________

169 **DAY FOUR**

1. How could you get from ☆ to Point 1, then from 1 to 2, and so on, until you reach Point 6? Use the words **right, left, up,** and **down**.

2. Draw a picture of your path.

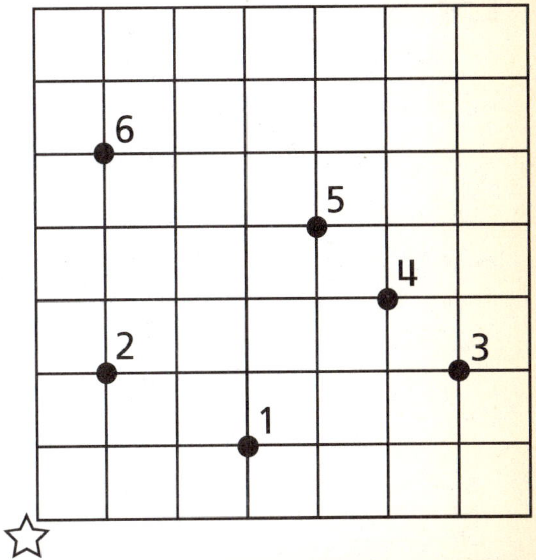

170 **DAY FIVE**

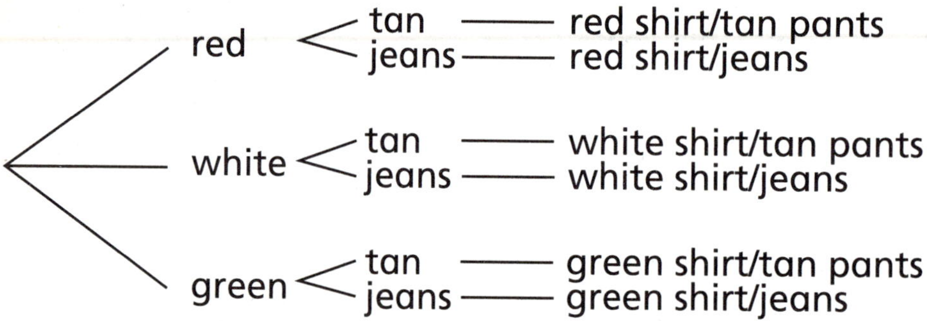

1. What does the tree diagram show?
2. What labels could you put above each of the three columns?

Name ______________________________

171 **DAY ONE**

Find each product. Then write the products in order from least to greatest.

1. 7×34
2. 3×68
3. 5×48
4. 9×27

172 **DAY TWO**

Gary is thinking of a 4-digit number.

- It has twice as many tens as thousands.
- The number of ones is equal to the sum of the hundreds and tens digits.
- It has a 3 in the thousands place.
- The number of hundreds is less than 2 and greater than 0.

What is the number?

Name ______________________________

173 **DAY THREE**

Which statements are reasonable?

A. Stefan walked 2 feet to school.

B. Angie's house is 1 mile long.

C. Mr. Lee is 6 feet tall.

D. Carla walked 1 mile in 20 minutes.

Name ______________________________

174 **DAY FOUR**

Draw an 8–by–8 grid like the one shown. Label START on the lower left corner of your grid. Follow the directions in order. Draw your path, making a dot on the grid after each set of directions.

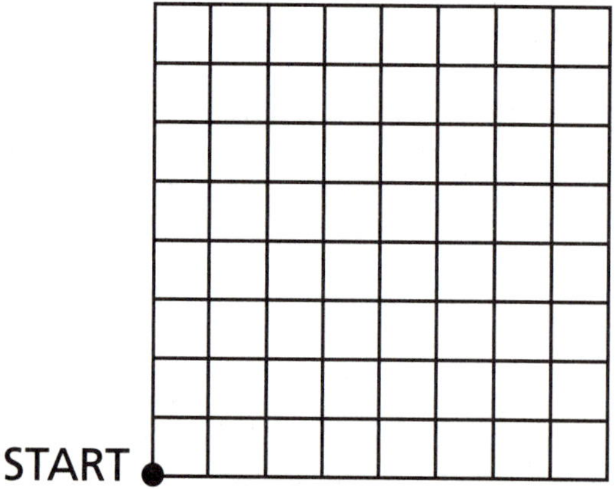

Directions:

- right 2, up 8
- right 6, down 8
- left 5, up 7
- right 4, down 2
- left 3, down 2
- right 3, down 2

175 **DAY FIVE**

Put the numbers where they belong in the Venn diagram.

8 10 18 20 24 30 60

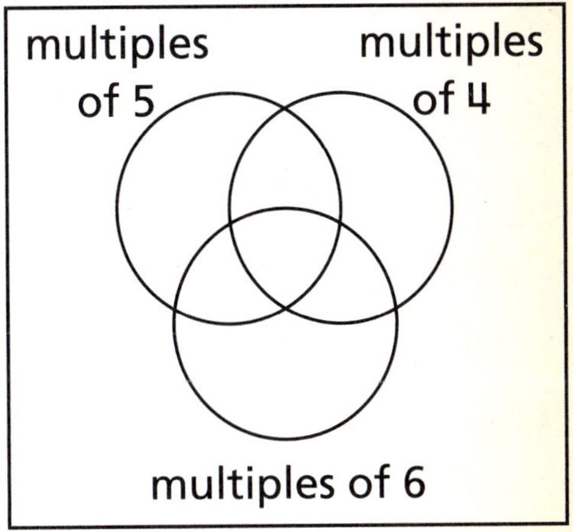

Name ______________________________ WEEK **36**

176 **DAY ONE**

Complete the table.

Burgers	1	2	3	4	5	6	7
Cost	\$1.05	\$2.10	\$3.15				

Name ____

177 **DAY TWO**

Write >, <, or = in the ◯ to make each true.

1. 0.40 ◯ 0.4

2. 0.32 ◯ 0.10

3. 0.09 ◯ 0.90

4. 0.71 ◯ 0.78

178 **DAY THREE**

Match each division problem with its quotient.

1. 65 ÷ 7 **a.** 9 R2

2. 78 ÷ 4 **b.** 11 R6

3. 57 ÷ 3 **c.** 15 R2

4. 92 ÷ 6 **d.** 19

5. 83 ÷ 7 **e.** 19 R2

Name ______________________________

179 **DAY FOUR**

Write the fraction shown by the shaded part of each figure.

1. 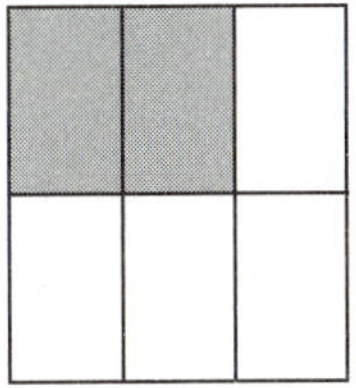

2. 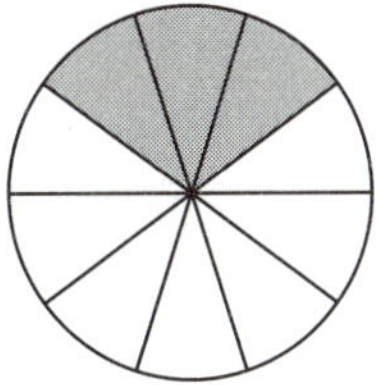

3. 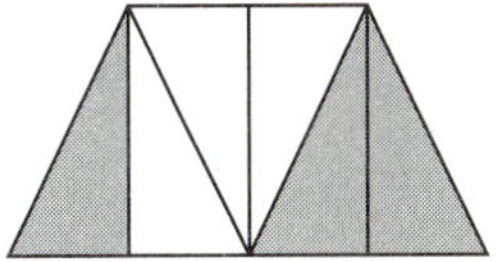

4. 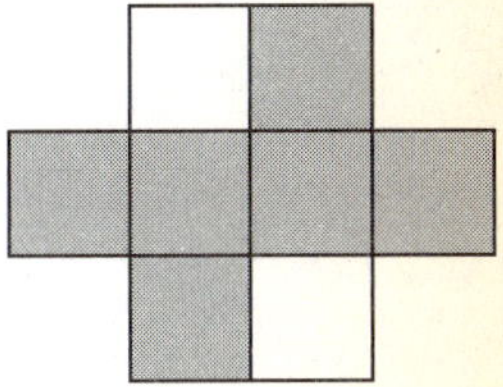

180 **DAY FIVE**

Draw the next two clock faces in the pattern.

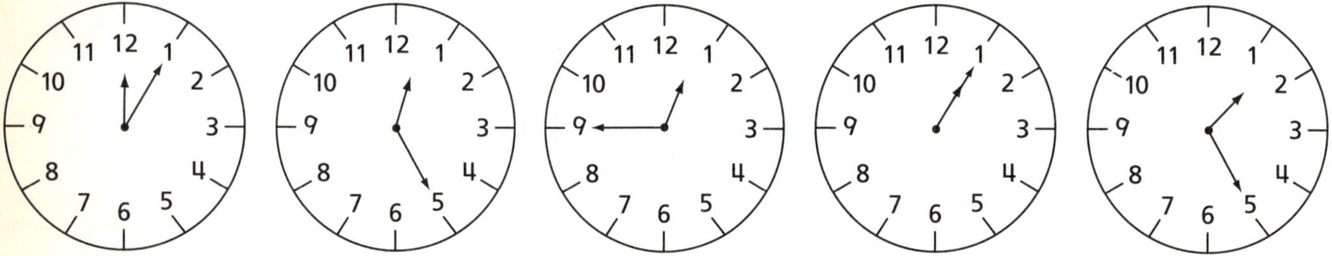